Gessica Xavier Torres
Ruth Abreu Araújo
Sheila Xavier Novais

# Neem extracts (Azadirachta indica A. Juss) in seed treatment

Gessica Xavier Torres
Ruth Abreu Araújo
Sheila Xavier Novais

# Neem extracts (Azadirachta indica A. Juss) in seed treatment

## The use of neem extracts in the treatment of vinegar seeds (Hibiscus sabdariffae)

**Imprint**
Any brand names and product names mentioned in this book are subject to trademark, brand or patent protection and are trademarks or registered trademarks of their respective holders. The use of brand names, product names, common names, trade names, product descriptions etc. even without a particular marking in this work is in no way to be construed to mean that such names may be regarded as unrestricted in respect of trademark and brand protection legislation and could thus be used by anyone.

Cover image: www.ingimage.com

This book is a translation from the original published under ISBN 978-613-9-72575-5.

Publisher:
Sciencia Scripts
is a trademark of
Dodo Books Indian Ocean Ltd. and OmniScriptum S.R.L publishing group

120 High Road, East Finchley, London, N2 9ED, United Kingdom
Str. Armeneasca 28/1, office 1, Chisinau MD-2012, Republic of Moldova, Europe
Printed at: see last page
ISBN: 978-620-7-90542-3

# Summary

# CHAPTER 1

# INTRODUCTION

Vegetables form the fundamental basis of human nutrition and in the universe of vegetables and fruit, many stand out for offering important components for humans to fulfil their basic functions, such as nutrients like vitamins, minerals and other bioactive substances that are essential for a balanced diet (CAMPOS et aL, 2006).

Planted from north to south in Brazil, the vinagreira (Hibiscus *sabdariffae* L.), also called roselle, sour okra, azedinha and quiabo-de-angola, stands out in Maranhão as the basis of local culinary dishes. This multi-purpose plant has its main use as a vegetable in Brazil in the consumption of its leaves, which are rich in vitamins A and B1, mineral salts and amino acids. The fruit and calyx are the raw material for making juices, jams and jellies. In the United States, it is common to make a type of wine from the juice extracted from the calyxes and leaves. In Asia (especially Japan and India) and Russia, there is a strong demand for dried vine flowers to make tea, either pure or mixed with traditional tea made from *Camellia sinensis* leaves, resulting in the popular "hibiscus tea". Some varieties are used to produce fibres for the textile industry. It also has potential as an ornamental plant. Finally, as a medicinal plant, its leaves are used as a fortifier and stomach stimulant and the flowers as antibacterial and antifungal (CARDOSO, 1997).

Given the nutritional potential of vinegar, which is still almost unknown in some Brazilian regions, the crop faces serious phytosanitary problems. The main diseases related to the crop are rot caused by *Phytophthora sp.* In regions with low relative humidity, the incidence of *Oidium sp. is* common. A new pathogen has also been reported in studies carried out in the state of Pará, where the presence of *Corynespora cassiicola* Berk. M.A. Curtis. The fungus is the causal agent of target spot on several important agricultural, forestry and ornamental species in Brazil and worldwide (POLTRONIERI, 2012).

There is still a great deal of concern when it comes to the crop's propagative material, as pathogens that attack seeds cause serious damage to the grapevine, making it unviable in terms of sanitary quality, especially during the storage phase. Amongst the main fungi that can be carried by seeds in storage we can mention *Aspergillus niger, A. flavus, A. glaucus, A. candidus, Penicillium spp, Rhizopus stolonifer, Phomopsis spp, Chaetomium sp, Colletotrichum sp.* Goulart (1997) states that to date, numerous fungi have been identified on soya beans, but only a few are worth highlighting because they are economically important, as listed below: *Phomopsis spp.* (anamorph of *Diaporthe spp.*), *Colletotrichum truncatum, Fusarium spp.* (mainly *F. semitectum), Aspergillus spp.* (mainly *A. flavus), Penicillium spp., Alternaria spp., Chaetomium sp., Rhizoctonia solani, Rhizopus stolonifer,* among other fungi.

The importance of the occurrence of pathogens in seeds can be assessed from different perspectives, all culminating in the economic dimension that each pathogen-host interaction determines. The economic significance of this association can be estimated based on the very expression of each disease as it occurs in nature, and considering the fact that the majority of known diseases can have their aetiological agents transmitted effectively by the seeds of their hosts (MACHADO, 2000a).

Studies related to seed treatment have been carried out over the years, and alternative methods have been used in the sanitary management of these seeds, among them the plant extract of some species combined with various forms of extraction.

Among these plants, Nim *(Azadiaachta indica* A .Juss) is an interesting alternative as it has shown results in reducing phytopathogens in both the aerial part and the soil. Different Nim products have been tested to control phytopathogens, with promising results on phytonematodes (CARNEIRO, 2002a). The fungitoxic effect of this plant has been tested on entomopathogenic fungi and in pathosystems involving soil fungi.

Considering the economic and social importance of vinegar and the lack of information on the crop's phytosanitary problems, the aim of this study was to assess the incidence of phytopathogenic fungi under the effect of different neem extracts in the treatment of vinegar seeds.

# CHAPTER 2

# LITERATURE REVIEW

## 2.1 Growing vinegar (*H/b/scus* *sabdariffae* L.)

### 2.1.1 History

*Hibiscus sabdariffa* L. (Malvaceae) is an important medicinal plant that originated in India, Sudan and Malaysia and was later taken to Africa, Southeast Asia and Central America. It is also known as "azedinha, azeda-da-guiné, caruru-azedo, caruruda-guiné, cha-da-jamaica, pampolha, pampulha, papoula, poppy-of-two-colours, quiabeiro-azedo, quiabo-azedo, quiabo-de-angola, quiabo-rosa, quiabo-roxo, rosélia and vinagreira" (MUKHTAR, 2007).

It was probably introduced to Brazil by Africans during the slave trade (BRASIL, 2010a).

In Maranhão, vinagreira is one of the most popular vegetables and is used to prepare typical dishes. Its leaves are rich in vitamins A and B, mineral salts and amino acids. In folk medicine it is used as a diuretic, calming and antiscorbutic. Vinagreira also has enormous potential as a source of colouring for the food industry and as a producer of fibres for the manufacture of burlap.

## 2.1.  2Features

A plant from the Malvaceae family, it is a vigorous annual shrub that can reach up to 3 metres in height, with a green or reddish stem. However, in order to use the plant, it must be managed by cutting it down to a height of around 1 to 2 metres. Its leaves are alternate, lobed and toothed and green or purple in colour. The flowers are yellowish-white, pink or purple, with red or white fleshy calyxes (Figure 1) that will form the fruit, small capsules (BRASIL, 2010b). In traditional medicine, it is used as a diuretic, to treat gastrointestinal disorders, liver infections, fever and hypertension (MONROY-ORTIZ & CASTILLOESPANA, 2007).

**Figure** 1-Vinagrape

**Source:** ervasnaturais.wordpress.com

It is the most popular leafy vegetable in the interior of Maranhão, prepared in a regional dish called cuxá. In the south-eastern states, as well as northern Paraná, where it is called roselle or gooseberry, its calyxes are used for juices and jellies. The Japanese colony, who call it umê, use the calyxes for pickles. It is also found in Pará, especially in Belém, where it is an obligatory product at the Mercado do Ver - o - Peso (KHATOUNIAN, 1994a).

Martins (1985) reports that from 1978 to 1983, in the São Luis wholesale market, it was the first vegetable in production in Maranhão and the eleventh in terms of quantity sold overall.

The domestication of the vine is believed to have been primitively linked to the utilisation of the seeds as food (MARTINS, 1985). Its composition includes 25.2 per cent crude protein, 21 per cent lipids and only traces of gossypol (AL-WANDAWI et al., 1984).

A 1975 IBGE survey found 227 cases of vinagreira consumption in Brazil, of which 82 were in non-metropolitan urban regions in the Northeast, 111 in the rural Northeast and 20 in the Belém metropolitan region. However, *on-site* observations made by Khatounian (1994b) suggest that most of the cases found in the Northeast occurred in Maranhão, since the species is infrequent or even absent in other states in the region.

In addition to its versatile use, the plant has some desirable agronomic characteristics. It grows in moderately fertile soils, does not require irrigation and is apparently not affected by gall nematodes. These attributes, combined with its long cycle, place it in a privileged position among summer leafy crops (KHATOUNIAN, 1994c).

With little work and on land unsuitable for other leafy greens, it produces an abundance of leaves over a long period of time. In contrast to other leafy greens eaten cooked, whose flavour is not very pronounced, the vinagreira has a distinct acidic taste, capable of breaking up the monotony of diets with little variety. This rare set of attributes, which combines culinary versatility, nutritional value and ease of cultivation, makes the vinagreira one of the plants with the greatest potential in the tropics (KHATOUNIAN, 1994c).

## 2.1.2 Economic importance

The most studied parts of the vine are the leaves and calyx. Red in colour and sour in taste, they are attracting the attention of the food and pharmaceutical industries, which are beginning to see the possibility of rationally exploiting this plant as a raw material for food production and as a natural source of colouring agents, thus demonstrating great economic potential. The calyxes can be used to decorate dishes, such as salads with a high antioxidant value, or in the preparation of jams, sweets, juices, syrups, jellies, wine, vinegar, sauces or eaten fresh (EMBRAPA, 2011).

This part of the plant is widely used in folk medicine, for example, the calyxes of Hibiscus sabdariffa L have long been used in folk medicine to treat high blood pressure, gaining wide acceptance in the treatment of many diseases throughout most of Brazil (EMBRAPA, 2011).

There are reports in popular medicine of the wide-ranging use of H. sabdariffa L as an antiseptic, aphrodisiac, astringent, digestive, diuretic and stomachic (AKINDAHUNSI; OLALEYE, 2003).

Hibiscus sabdariffa L is a popular remedy for abscesses, heart disease and hypertension. Hibiscus tea has been shown to be effective in reducing blood pressure in patients with hypertension (AKINDAHUNSI; OLALEYE, 2003).

In southern Mexico, a popular drink is made from the dried calyxes of the plant, which is also traditionally used by the population to treat obesity (EMBRAPA, 2011).

Similarly, in Nigeria, there is a high consumption of a drink made from the calyxes of Hibiscus sabdariffa L. Studies have proven the use of H. sabdariffa L as a diuretic, uricosuric, antimicrobial, mild laxative, sedative, antihypertensive and antitussive agent, as well as to reduce total lipid, cholesterol and triglyceride levels, in gastrointestinal and kidney stone treatment, and to treat liver damage and the effects of drunkenness. More recently, there are indications that H. sabdariffa L acts as an antioxidant, antimutagenic, antitumour and antileukaemic agent (EMBRAPA, 2011).

## 2.2 Importance of plant diseases

According to Michereff (2001), plant diseases are important to humans because they cause damage to plants and their products, as well as directly or indirectly influencing the profitability of the business.
agricultural enterprise. He says that plant diseases can cause serious damage, which is explained below.

- Plant diseases can limit the types or varieties of plants that can grow in a given geographical area. Plant diseases can also determine the type of agricultural industry and the level of unemployment in a given area by influencing the type and quantity of products available for processing and packaging. On the other hand, plant diseases are responsible for the creation of new industries that produce agrochemicals, machinery and develop methods for disease control.

- Plant diseases reduce the quantity and quality of plant products, and losses vary depending on the species of plant or the products obtained from it, the pathogen, the location, the environment, the control measures adopted and the combination of these factors. The amount of losses varies from minimal percentages to 100 per cent. Plants and/or their products can be reduced quantitatively due to diseases in the field, as is the case with most plant diseases, or by diseases during storage, as is the case with fruit, vegetable and seed rots. The decline in the quality of plant products often results in notable losses. For example, spots, scab, pustules and other infections on fruit, vegetables and ornamental plants may have very little effect on the quantity produced, but the inferior quality of the product can drastically reduce its market value and may even result in a total loss.

☐ Plant diseases can make plants poisonous to humans and animals. Some diseases, such as rye and wheat spur caused by the fungus *Claviceps purpurea,* can make plant products unsuitable for human or animal consumption by contaminating them with poisonous fruiting structures. Many grains and some other seeds are often contaminated or infected with one or more fungi that produce highly toxic compounds known as mycotoxins. Consumption of these products by humans or animals can lead to the development of serious diseases of internal organs and the nervous system.

☐ Plant diseases can cause economic losses. Farmers can incur financial losses due to plant diseases when they need to produce disease-resistant varieties or plant species that are less productive, more costly or less commercially acceptable than other varieties, as well as when they need to spray or adopt other disease control measures, resulting in expenditure on pesticides, machinery, product storage space and labour. When products are plentiful and prices are low, the limited time that these products can be kept fresh and healthy means that they need to be sold in a short space of time. Separating healthy and diseased plant products to avoid spreading the disease also increases marketing costs.

## 2.3  Seed pathology

### 2.3.1 Historical evolution

The association of pathogens with seeds, according to Baker and Smith (1966), dates back more than eight centuries. Baker (1979) admits that the development of mechanisms for transmitting pathogens via seeds probably began around the time that angiosperms became the dominant flora on Earth and seeds became the usual way of multiplying plants. In historical terms, many of the events that led to the emergence of Seed Pathology were, over many decades, dealt with by General Phytopathology, making these facts an integral part of the very history of this Science. The first concrete report on the association of a pathogen with seeds was made by Hellwing in 1699, according to Baker (1972). In this publication, reference was made to the transport of *Claviceps purpurea* sclerotia along with rye seed (MACHADO, 1988a).

## 2.3.2 Pathogen association with the seed

It is considered extremely important that the damage caused by the association of pathogens with seeds is not limited to direct losses of plant populations in the field, but also includes a series of other implications which, in a more accentuated form, can lead to irreparable damage to the entire agricultural system. For a large number of devastating diseases, the seeds carrying their causal agents are the only way they can spread and perpetuate themselves in nature (MACHADO, 2000b).

Machado (2000c) cites reports of the association of pathogens with seeds that have been published in periodic compilations by the International Seed Testing Association (ISTA). In these publications, the association of approximately 1500 species of pathogens with their hosts is reported. This number has certainly increased as the relationship between pathogens and their hosts has become better understood and the natural co-evolution of this biological interaction has been taken into account.

Among the pathogens that can associate with plant seeds, fungi form the largest group, followed by bacteria and, to a lesser extent, viruses and nematodes. Figure 2 shows some of the damaging implications of pathogen-seed interaction, emphasising that, depending on the view of each segment of the production system, each of these implications can have a different meaning.

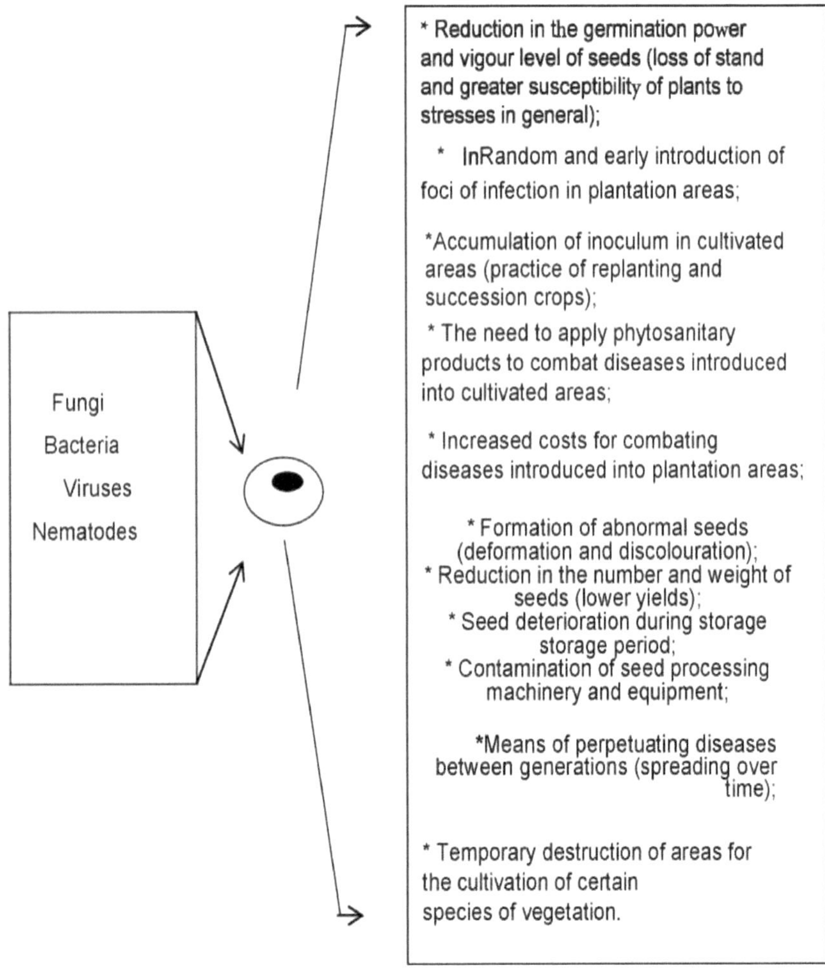



**Figure 2 -** Schematic representation of the implications of the association of pathogens with seeds.

## 2.4 Importance of storage fungi

Storage fungi" are the main culprits in the loss of viability of seeds stored with a moisture content above the critical value. It has been shown in various experiments that most of these fungi preferentially attack the embryo and some only the embryo. Samples of wheat, barley, maize and sorghum seeds were stored at temperatures and humidity favourable to the growth of "storage fungi", but free of inoculum and maintained their viability for several months. Samples stored under the same conditions, but carrying inoculum, lost their viability within a few months (ONKAR, 1985). Lopez and Christensen (1967) stored maize seeds with 19% moisture at 20-250C. One part of the samples was inoculated with A. *flavus* and the other remained free of the fungus. After 74 days, the "healthy" samples had maintained 97% germination and those inoculated with *A. flavus had* only 14% germination.

The most important factors in determining a storage fungus infection in seeds are: seed moisture content, relative humidity of the environment, temperature and time. These factors are interrelated and should be considered complex. For storage to be successful, each factor must be known in its own right and together. Humidity is a major factor, since the moisture contained in the seeds establishes such a relative humidity around them that it can favour fungal growth. The humidity of the environment in turn determines the moisture content of the seeds when in equilibrium (ONKAR, 1985).

## 2.5 Seed treatment for pathogen control

Machado (2000d) states that in the integrated control of plant diseases, seed treatment is a valuable measure due to its simplicity of execution, relative low cost and effectiveness in various aspects. The fact that it controls diseases in the phase prior to planting a crop or at the time of sowing makes seed treatment one of the most recommended measures in modern agriculture, enabling less use of chemical pesticides and, consequently, avoiding serious environmental pollution problems Generally speaking, seed treatment to control pathogens can be practised using various resources, which are based on research knowledge that derives from the action or interference of factors directly on pathogens or on the diseases that these agents cause. He also states that three methods can be used in the practice of seed treatment:

a) Chemical, which consists of incorporating artificially developed chemical products into the seeds;

b) Physical, which involves exposing the seeds to heat or another controlled physical agent;

c) Biological, which is based on the incorporation of antagonistic organisms into the seeds.

Combining these resources has been a recommended tactic for controlling numerous pathogens. The reason for the existence of different resources for seed treatment is based on the fact that the diversity and nature of the agents that cause diseases are enormous and not always does a single treatment method provide control in all cases. Various controllable and non-controllable factors can also interfere with the effectiveness of a treatment method (MACHADO, 2000d).

## 2.6 Alternative seed treatment

## 2.6.1 Plant extracts for disease control

A new agricultural policy through alternative agriculture has been developed, aimed at minimising impacts on the environment and humans through alternative plant disease control, which includes biological control, induction of resistance in plants (BETTIOL, 1991) and the use of alternative products to chemical control, such as plant extracts and oils, either through their direct fungitoxic action or indirectly through the activation of defence mechanisms in the treated crops.

One plant species that has been widely used and studied, with promising results, is garlic (A//um *sativum).* The plant contains two substances, alliinase and alliin, which are stored separately and, when their membranes are broken, form allicin, which is responsible for the plant's defence. Its toxic effects inactivate micro-organisms (HEINZMANN, 2001). According to Ranasinghe et al. (2002), cloves (Syzygium *aromaticum)* and cinnamon *(Cinnamomum zey/anicum)* also have fungitoxic properties. The authors state that the eugenol present in cloves may be the toxic component, both in the aqueous extract and in the essential oil. Cinnamon bark, on the other hand, has cinnamaldehyde as its main antimicrobial constituent. These studies have often shown differences in the inhibition of the development of phytopathogens in relation to the different concentrations of plant extracts tested. In general, there is a tendency for the higher the concentration in the culture medium, the greater the effect of this extract in terms of inhibiting the mycelial growth of the fungi. Motoyama et al. (2003a), when working with citrus extracts, observed inhibition of the mycelial growth of *Fusarium semitectum from a* concentration of 100 ppm, however, the maximum inhibition was seen when using the highest concentration, 5,000 ppm (48.18%). Barros et al. (1995), in experiments with the fungi of the genera *Curvularia* and *Alternaria,* found greater inhibition of the development of fungal colonies with the use of garlic extract at a concentration of 10,000 ppm when compared to the concentration of 1,000 ppm.

On other occasions, the effects of the concentrations of the extracts were not detected. Rodrigues et al. (2006), evaluating the potential of the *Ocimum gratissimum* plant to control *Bipolaris sorokiniana,* found that both the extract sterilised by filtration and autoclaved, at concentrations of 5, 10, 15, 20 and 50% and 10, 15, 20 and 50%, respectively, provided similar percentages of inhibition of the pathogen's mycelial growth.

*Lemon* balm *(Lippia alba)* also has antifungal activity. According to Schwan Estrada *et al.* (2000), both the crude extract and the essential oil of lemon balm have been used for *in vitro* studies to inhibit the mycelial growth and sporulation of phytopathogenic fungi *(Rhizoctonia solani, Sclerotium rolfsii, Alternaria alternata, Phytophthora sp. and Colletotrichum graminicola) and* in bioassays for the induction of phytoalexins in sorghum (deoxyanthocyanidins) and soya (gliceolin).

## 2.6.2 General characteristics of neem

Neem is a plant of Asian origin, belonging to the Meliaceae family, native to Burma and the arid regions of India (SAXENA, 1983). Neem (Azadirachta indica A.

Juss) can also be found under the names neen, margosa, nime, Indian lila, or even Melia azadirachta L., Melia indica (A. Juss.) Brandis and Antelaea azadirachta (L.) Adelb. (KOULetaL, 1990).

Neem can withstand severe drought and high temperatures, but is very sensitive to cold. In Brazil, whatever the purpose of planting, all areas where the average annual temperature is below 20 °C are unsuitable for growing neem. When the objective is only to produce leaves, locations with an average annual temperature of 20 °C to 21 °C can provide satisfactory results, as long as the average temperature of the coldest month is e" 16.0 °C.

Whatever the objective, areas where the average annual temperature is between 21 °C and 23 °C and e" 23 °C, respectively, are considered good and optimal for growing neem. The species can be grown in places with different rainfall regimes. There are successful plantations for fruit production from the region of Petrolina, PE,/ Juazeiro, BA, with average rainfall of 600 mm/year and seven months of drought, to the west of the state of São Paulo, with rainfall of around 1,200-1,400mm and 3-4 months with little rain. It should be remembered that, worldwide, neem is a particularly valuable plant for cultivation in sub-humid and semi-arid tropical regions. In the north-east of Pará, where rainfall is over 2,000 mm/year and there is no water deficit, plantations grow well, but there is no reliable information on fruit production, the most noble purpose of neem.

In the South, the favourable climate is only found in part of northern Paraná and mainly in the north-west (Caiuá Sandstone). The Southeast, Centre-West, Northeast and part of the North (where the climate is Awi, according to the Kõeppen classification) are the regions with the most suitable climatic conditions for growing neem. In the hottest and driest locations, the species produces good seed for oil extraction.

In Brazil, in the more humid biomes (Cerrado, Atlantic Forest and Amazon), the vast majority of neem plantations are generally located on deep, drained soils: Argissolos, Quartzarenic Neossolos and Latossolos. In the Caatinga and its transition to the Zona da Mata, there are good plantations established on shallow soils, such as Luvissolos and Neossolos Litólicos. The most suitable soils are those with a pH between 5.0 and 7.0 with low levels of exchangeable aluminium, high levels of exchangeable bases and high base saturation. These soils are naturally found only in the Northeast.

To grow neem in acidic soils (pH < 5.0), it is necessary to correct the acidity by liming. The species is demanding of N, P, K and Ca. Depending on the region, the physical aspects of the soil can be more limiting than the chemical ones. The species can withstand long periods of drought, but does not tolerate waterlogged soils, even temporarily.

It is not picky about deep soils, but it does require permanently drained or well-drained soils: hence, the soils in the wetter biomes (Cerrado, Atlantic Forest and Amazon) must necessarily be deep and aerated, while in the Caatinga they can be shallow. Based on field observations, in the wetter biomes
(Cerrado, Atlantic Forest and Amazon), the ideal condition is that the water table is permanently at least 2 metres from the surface. Valley bottoms prone to water accumulation should be avoided for planting.

## 2.6.3 Growing neem to extract vegetable oil

The most valuable neem products come from the fruit: the oil from the seeds and the cake, which is the residue from obtaining the oil. The economic viability of plantations to produce the fruit depends on the existence of processing industries that consume it.

In turn, the operation of a neem fruit milling plant requires a minimum amount of raw material in its vicinity: therefore, in a micro-region, a minimum planted area is needed to make the entire chain possible. In this sense, the following explanations should be taken as a basis for orientating undertakings.

## 2.6.2 Neem      extract

Various natural products, including medicinal plant extracts, have shown their ability to control plant diseases, both through their direct antimicrobial activity and indirectly by inducing resistance (MOTOYAMA et al., 2003b).

Plants are rich in secondary metabolites such as tannins, terpenoids, alkaloids and flavonoids (COWAN, 1999). These compounds can have antifungal activity and have been studied for the development of phytosanitary products.

Amongst the plants with antifungal potential, Indian neem *(Azadrrachta indica* A. Juss), also known as *Melia indica* Brandis, stands out. The plant belongs to the Meliaceae family and its insecticidal and medicinal properties have been known to the Indian population for centuries (PADOLE; THAKKAR, 2009a). It comes from Asia and is cultivated in several African countries, Australia and Brazil, where it has adapted well to the Southeast, Centre-West and Northeast regions (NEVES, 1996).

The chemical nature of neem has been studied since 1919 and various substances have been described. In 1942, three substances were identified: nimbim, nimbidim and nimbinene (PADOLE; THAKKAR, 2009b). In the 1970s and 1980s, more than 150 compounds isolated from leaves, branches and seeds were identified, the most active of which belong to the limonoid class (SCHMUTTERER, 1990; cited by VIANA et al, 2006a).

Among the isolated substances, azadirachtin is the most studied compound in neem. It is characterised as a very complex terpene and its molecule has not yet been synthesised, so all the products are prepared by extracting the compounds from the plant (MARTINEZ, 2002a). Azadirachtin is sensitive to photodegradation and its insecticidal action can be reduced by ultraviolet rays; however, results obtained in the field have indicated that Neem extracts applied to crops can remain insecticidal for approximately seven days (SCHMUTTERER, 1990; cited by MARTINEZ, 2002b).

Studies carried out at Embrapa Milho e Sorgo by Viana et al. (2006b) indicate that the azadirachtin content in neem leaves varies according to the time of year. Although azadirachtin is the most studied substance in the neem plant due to its insecticidal effect, other compounds in the plant have been described with effects on phytopathogens. Govindachari et al. (1998), studying the antifungal activity of terpenoids that make up neem oil, found that azadirachtin did not affect the growth of three phytopathogenic fungi: *Drechslera oryzae* (Breda de Haan) Subram; Jain, *Fusarium oxysporum and Alternaria tenuis* Nees; while salanin, nimbin, epoxyazadiradione, deacetylnimbin and azadiradione showed control. These five terpenoids were effective in controlling fungi and showed greater action when applied in a mixture than when tested alone, suggesting a synergistic effect of these substances.

According to Carneiro (2002b), different neem products have been tested *in vitro to* control phytopathogens and in greenhouse conditions with direct application of the product to the plant. Nim cake, a by-product of extracting the oil from its seeds, contains a large number of triterpenoids and is rich in amino acids, making it a potential organic fertiliser rich in nutrients such as nitrogen, phosphorus and potassium (PADOLE; THAKKAR 2009c). The antimicrobial effect of neem seed extracts has been studied and found to be effective against a range of pathogenic bacteria and fungi (COVENTRY; ALLAN, 2001). Moslem; El-Kholie (2009) evaluated the antifungal effect of 40 per cent neem seed and leaf extracts on phytopathogenic fungi. The authors found a 100 per cent reduction in the development of *Fusarium oxysporum* and *Rhizoctonia solani* Kuhn, and an 80.7 per cent and 71.2 per cent reduction in *Alternaria solani* (Ellis; G. Martin) L.R. Jones; Grout, and *Sclerotinia sclerotiorum* (Lib.) de Bary, respectively.

*In vitro* tests revealed efficiency in reducing the development of *Curvularia lunata* (Wakker) Boedijn, a fungus that causes leaf spots on some crops. Neem leaf extract added to the culture medium reduced the diameter of the colony by 73%, the dry weight of the mycelium by 83% and the percentage of spore germination by 94% (BHOWMICK; VARDHAN, 1981; CITED BY CARNEIRO, 2002c).

Studies carried out by Hassanein et al. (2008) showed that ethanolic extracts of neem leaves were 50.44% and 100% effective in inhibiting the growth of *Alternaria solani* and *Fusarium oxysporum,* respectively. Similar results were recorded by Ashraf; Javaid (2007) who, when evaluating Neem leaf extract on *Macrophomina phaseolina (*Tassi) Goid, obtained an 85 % reduction in fungal biomass. The effect of aqueous extracts of neem leaves *(Azadiracrta indica),* jujube *(Ziziphus spina-chisti* (L.) and *Zygophillum coccineum chisti* (L.) at a concentration of 25 per cent were also tested and showed a reduction in the development of *Fusarium solani of* 82.28; 79.15

and 76.95 per cent for the respective extracts (HAIKAL, 2007).

The incorporation of neem leaves into the soil in a greenhouse was evaluated by Silva; Pereira (2008) to control the *Fusarium* x *Meloidogyne* complex in okra. The authors reported that incorporating 5% neem leaves into the soil was effective in controlling *Fusarium oxysporum* f. sp. *vasinfectum \MC.* Snyder; H.N. Hansen and *Meloidogyne incognita* (Kofoid; White) Chitwood race 1. The application of plant extracts associated with the biological control agent, *Trichoderma harzianum* Rifai, to control soil fungi was investigated by Haikal (2007).

The author evaluated the effect of plant extracts, including *Azadirachta indica* L., at concentrations of 5, 10, 15, 20 and 25% to control root rot in cucumbers caused by *Fusarium solani* and observed that all the concentrations tested were efficient in reducing fungal biomass. He also found that cucumber seedlings irrigated with the extract at a concentration of 25% in combination with the application of *T. harzianum* showed disease rates of 2.44% compared to the control which showed an incidence of 41.46%.

In addition to studies on the direct effect of neem leaf extracts on phytopathogens, the literature also contains reports on its effect as a resistance inducer in plants. *Sesame* plants *(Sesamo indicum* L. : Syn. S. L. *orientale)* that received an application of neem extract had a 75% reduction in the incidence of alternaria leaf spot compared to the control treatment and showed significantly higher phenylalanine ammonia lyase activity than the control. In addition, the level of peroxidase was ten times higher in the samples that received the extract (GULERIA; KUMAR, 2006).

## 2.7 Main forms of plant extraction

According to Food Ingredientes Brasil (2010), extracts are concentrated preparations, of various possible consistencies, obtained from plant raw materials that have or have not undergone prior treatment (enzymatic inactivation, grinding, etc.) and prepared by processes involving a solvent. This basically involves two stages in the manufacturing process: the separation of specific compounds from a complex medium (the drug, or part of the plant used, root, stem, leaf) with the use of a solvent; and concentration, by more or less complete elimination of solvents. There are a number of extraction technologies, the most commonly used of which are:

- Maceration: consists of the plant drug simply coming into contact with the extracting liquid for a set period of time;

- Infusion: boiling water is added to the plant;

- Decoction: the water in contact with the plant comes to the boil;

- Digestion: the drug-solvent contact is maintained at a temperature of 40 to 60 degrees Celcius;
  - Percolation: this is undoubtedly the process that, due to its dynamics and possible tricks, allows for greater extraction, more efficient extraction. Passing the extracting liquid through the ground drug in devices known as percolators, controlling the flow and varying the mixture of extracting solvents, optimises the process;

- Distillation: process in which the plant, in contact with water or alcohol, is subjected to distillation;
- Drying: when the liquid extract has had its solvent removed, it can be done by simple heating and evaporation or subjected to *spray dryer, drum dryer,* vacuum evaporation and concentration, membrane concentration and other processes.

# CHAPTER 3

# MATERIALS AND METHODS

## 3.1 Place of the experiment

The experiments were conducted at the Phytopathology Laboratory of the State University of Maranhão - UEMA, Paulo VI campus, São Luís, between July and August.

## 3.2 Seed health using the filter paper method ("blotter test")

The vinegar seed samples were first disinfected for five minutes by immersing them in a solution of sodium hypochlorite (NaCIO) at 1.5 % active chlorine, followed by two washes with sterilised water.

The seeds were then plated in Petri dishes, previously disinfected by autoclaving at 1 atm for 20 minutes and by exposure to ultraviolet (UV) light for 20 minutes, containing three layers of sterilised filter paper moistened with sterilised distilled water. 400 seeds were plated according to the pre-established seed analysis rules (BRASIL, 2009), using 20 seeds per container. The seeds were incubated under 12-hour photoperiod conditions at a temperature of approximately 26±5°C for seven days (PINTO, 2005). Population survey

The fungal colonies were analysed using a binocular magnifying glass (Zeiss, 50x magnification) seven days after plating. The colonies that developed on the seeds and seedlings were transferred to potato-dextrose-agar (BDA) medium for identification.

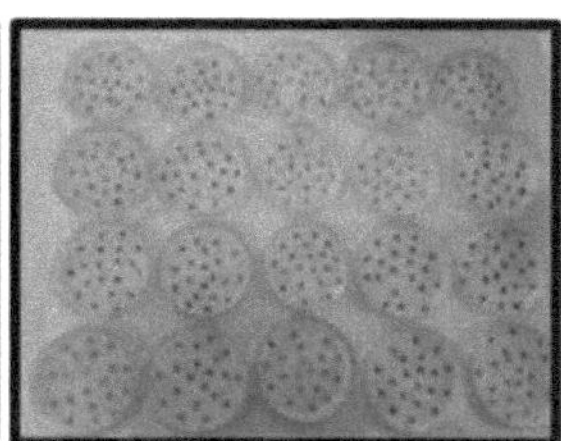

**Figure 3 -** Procedure for the seed health test: A (preparing the plates for the seed health test), B (selecting the vinegar seeds), C (seeds in the incubation process).

## 3.3 Obtaining the extracts

### 3.3.1 Hydro-alcoholic extract

Neem leaves were harvested and weighed (200g) in the morning at the State University of Maranhão - São Luís, then macerated and submerged in a hydro-alcoholic solution containing 500ml of 92.8% alcohol + 500ml of distilled water. The solution was collected in a beaker and left to rest for four days. Afterwards, the solution was filtered through cotton wool and taken to a bain marie at 50°C until it reached 700 ml so that 30% of the concentrated alcohol could evaporate, according to the methodology adopted by Cáceres et al (1995) (Figure 4).

**Figure 4 -** Obtaining the hydra alcohol extract: A (weighing the leaves), B (resting process for the hydra alcohol extract), C (evaporation of the alcohol in a bain marie).

### 3.3.2 Extract    by infusion

Neem leaves were collected and weighed (200 g) in the morning at the State University of Maranhão - São Luís, and washed in running distilled water. Distilled water, previously heated to 100°C, was added to a beaker containing the neem leaves. After this procedure, the leaves were left to rest for 15 minutes and then filtered through quantitative filter paper (Figure 5) under aseptic conditions in a laminar flow chamber.

**Figure 5** - obtaining the extract by infusion: A (weighing the leaves) B (filtering process for the extract by infusion).

## 3.2.1  Aqueous extract

30 g of neem leaves were weighed into 120 ml of distilled water, crushed in a blender and initially filtered through cotton wool, then filter paper was used for more precise filtering (Figure 6).

**Figure 6** - Preparation of the aqueous extract: A (crushing the leaves), B (filtering the extract), C (extract after filtering).

## 3.4 *In vitro* treatment of vinagrass seeds with different neem extracts

The vinegar seed samples were first disinfected for five minutes by immersing them in a solution of sodium hypochlorite (NaClO) at 1.5 % active chlorine, followed by two washes with sterilised water.

They were treated with Carbomax 500 SC fungicide, placed in a container and taken to a Vortex shaker for 10 minutes according to the manufacturer's recommendations, after which they were plated in Petri dishes (Figure 7).

The seeds that had been treated with hydroalcoholic, infusion and aqueous extracts were immersed for 30 minutes, dried (Figure 7), then plated in Petri dishes and incubated under a 12-hour photoperiod at a temperature of approximately 26±5°C for seven days. After the seven days of incubation, the fungi present on the seeds were quantified using a stereoscopic microscope (Zeiss at 50x magnification), which was used to visualise the individual colonies present on the seeds (Figure 8).

**Figure 7 - Seed** treatment process: A (seed treatment with hydro-alcoholic extract), B (seed treatment with aqueous extract), C (seed treatment with infusion extract), D (preparation of fungicide-based solution), E (seed treatment with Carbomax 500 SC fungicide).

The design was entirely randomised (DIC), with three extracts + one fungicide treatment + control, using five replicates, with 20 seeds per container, kept 1-2 cm apart. The data obtained was analysed using the Assistat software, updated version 7.7 beta, according to the Tukey test at 5% probability.

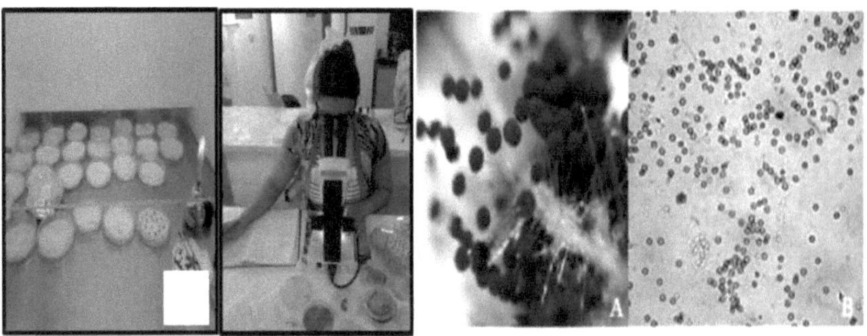

Figure 8 - Process of setting up and analysing the experiment: A (process of setting up the treatments) B (analysis of the seeds after 7 days of incubation). C (conidia and conidiophores of fungi of the genus *Aspergillus)*, D (conidia visualised on a slide). Source: www.infobibos.com

# CHAPTER 4

# RESULTS AND DISCUSSION

## 4.1 Seed health using the filter paper method (blotter test)

According to the results obtained in the health test (Table 1), the presence of fungi of great importance for stored seeds was observed, *namely Aspergillus flavus, Arpergillus niger, Rhizopus stolonifer, Chaetomium sp, Colletotrichum sp., Phomopsis sp., with* the fungus *Colletotrichum sp. having the* highest incidence, unlike the fungus *Phomopsis sp.,* which had the lowest incidence, with a percentage of 31.25% and 0.25%, respectively. There was a percentage of germinated seeds of 1.25% and non-germinated seeds of 98.75%, with a total percentage of healthy seeds of 43% and contaminated seeds of 57%.

Table 1 - Health test to assess fungal incidence in vinegar seeds *(Hibiscus sabdariffae)* and percentage of health and germination.

| Incidence (%) | | | | |
|---|---|---|---|---|
| Fungi | | | | |
| *Colletotrichum sp.* | *A. flavus* | *A. Niger* | *R. stolonifer* | *Chaetomium sp.* |
| Seeds | | | | |
| Germinated (%) | Non-germinated (%) | Healthy (%) | Contaminated (%) | |
| 1,25 | 98,75 | 43 | 57 | |

A germination percentage of only 5.5% was observed in the vinegar seeds, while the non-germinated seeds had a result of 94.5%. The percentage of healthy and contaminated seeds was 96.5% and 3.5%, respectively.

Similar results to this study were found by Lazarotto et al (2010a), in a health test with paineira seeds (Ce/ba *speciosa)* where they observed the presence of 13 genera of fungi including *Aspergillus sp., Colletotrichum sp., Rhizopus sp.,* among other potentially pathogenic fungi. In the results of Lazarotto et al (2010b) there was a high incidence of saprophytic fungi, with the most common genera being

*Penicillium* and *Rhizopus,* which appeared in all the samples. The *Rhizopus* fungus in this study was one of the least common, with a percentage of 4.25 %, contrasting with the results of the previous study.

Lazarotto et al (2010c) also stated that other fungi occurred in lower frequencies, with the genus *Aspergillus sp. standing* out, confirming the results of this study. This is considered a storage fungus, as its incidence increases with the post-harvest period, causing seed rot during germination and storage.

It can be seen that the presence of these fungi is constant, leaving us on alert as to how to store the seeds and where to keep them in a suitable state of health for future use, thus ensuring the longevity and viability of the seed over a long period of time.

## 4.2 *In vitro* seed treatment with neem extracts

The results of Table 2 in relation to incidence show that there was no significant difference between the treatments and the control in relation to the fungi *Colletotrichum sp. and Rhizopus stolonifer. The* latter only appeared in the aqueous treatment, which may have favoured the appearance of this fungus. According to Lopes et al (2011) in studies related to garlic extract (Al/um *sativum)* it was observed that the fungus of the genus *Rhizopus had a* 1% incidence in angico seeds *(Anadenantheaa colubrina).*

**Table 2 -** Averages of the treatments submitted to neem leaf extracts.

| T reatment | Incidence(%)* | | | |
| --- | --- | --- | --- | --- |
| | Fungi | | | |
| | *Colletotrichum sp.* | *A. flavus* | *A.niger* | *R. stolonifer* |
| Witness | 3,8 a | 0,4 b | 1,0 b | 0,0 a |
| Fungicide Carbomax | 1,4 a | 1.4 ab | 1,4 b | 0,0 a |
| Hydra alcohol extract | 2,8 a | 0,2 b | 0,0 b | 0,0 a |
| Aqueous extract | 3,2 a | 0,6 b | 0,2 b | 0,2 a |
| Extract by infusion | 1,6 a | 2,4 a | 4,4 a | 0,0 a |

However, when it comes to seeds, the *Rhizopus* genus is generally considered a contaminating fungus (REGO, 1995). This species can cause seed rot by attacking the cotyledons, rotting them before emergence when only the root system has been emitted, and then progressing to the seedling (MENTEN, 1995a). Some authors have found that certain fungi, when infecting seeds, damage their cellular structures in such a way as to jeopardise their viability and may even cause them to die (WETZEL, 1987; MACHADO, 1988b; MENTEN, 1995b).

With regard to the *A. flavus fungus,* there was a significant difference between the infusion extract treatment and the control treatment and the other treatments, which did not differ only from the fungicide Carbomax SC 500. As for A. *niger, there was a* significant difference between the infusion treatment and the other treatments.

It is worth noting that the incidence of all the fungi in the control was relatively low when compared to the treatments, which is why there was no significant difference.

Table 3 shows that the fungus *Colletotrichum sp.* obtained the highest control rates with the Carbomax fungicide treatment and the infusion extract treatment, with percentages of 63.16% and 57.89%, respectively. On the other hand, *A. flavus and A.* niger were 100% controlled by the Carbomax treatment. These results are similar to those of Silva et al (2008), who found that the extract of Brazilian boldo *(Plectranihus barbatus)* and other medicinal plants were able to inhibit the mycelial growth of phytopathogenic fungi of the *Colletotrichum* genus.

**Table 3 -** Percentage of fungal control of vinegar seeds treated with neem extracts.

| Treatment | Control (%) | | | |
|---|---|---|---|---|
| | Fungi | | | |
| | *Colletotrichum sp* | *A. flavus* | *A. Niger* | *R. stolonifer* |
| Witness | - | - | ■ | · |
| Fungicide | 63,16 | 100 | 100 | · |
| Hydro extract | 26,32 | 50 | 100 | · |
| alcoholic Aqueous extract | 15,79 | NC | 80 | NC |

| Extract by infusion | 57,89 | NC | NC | · |

Coutinho et aL, (1999) evaluated the efficiency of some extracts from plants in the Anacardiaceae family in the treatment of bean seeds *(Phaseolus vulgaris* L.) and found that when used alone, they reduced the occurrence of some fungi, especially *Aspergillus flavus;* however, their effects were more noticeable when they were mixed with chemical fungicides. In a study carried out by Botelho (2006), the fungicide captam reduced most of the fungi detected in ipê-roxo and ipê-amarelo seeds. This was also confirmed by Sales (1992) in the control of various fungal genera identified in ipê-amarelo and ipê-roxo seeds, such as *Alternaria, Phoma, Phomopsis and Fusarium.*

These results corroborate those found in this study. This is why chemical seed treatment is still the most widely used, because when compared to other forms of treatment there is a greater response, as was the case in the study carried out with vinegar seeds.

It is worth pointing out that the treatment with the hydro-alcoholic extract also managed to control the *A. niger* fungus by 100%. With regard to the treatments with aqueous extract and infusion, it was observed that there was no control of the fungi *A. flavus and Rhizopus stolonifer,* as the number of colonies found in the treatments was higher than that of the control. According to Lazarotto (201 Od), studies carried out on the treatment of seeds with hot water at 50°C showed that there was a greater variety of fungi present in this treatment, but the incidence of *Aspergillus niger* was very high, probably due to the use of heat and high humidity which favours the appearance of seed rotting fungi.The presence of holes in the vinegar seeds was observed, probably due to the appearance of an insect of the order Coleoptera. Better known as *Acanthoscelides obtectus* (bean beetle), it is a brownish primary insect that measures 3.5 to 4.5 mm in length (Figure 9). Bean beetles cause great damage to stored beans due to the presence of eggs in the beans, larval galleries, adult emergence holes, dead insects and droppings, which affect the quality of the product. Beans intended for sowing are also damaged because the embryo is destroyed (SARTORATO et al., 1987), which may have caused the seeds not to germinate.

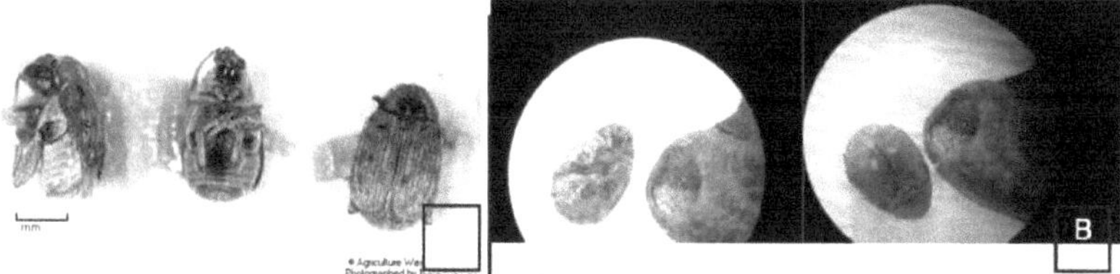

**Figure 9 -** Characteristics of the insect pest *Acanthoscelides obtectus: K* (visualisation from various angles of the insect's body), B (damage caused by the insect's attack on the vinegar seed seen with the aid of a stereoscopic microscope).

# CHAPTER 5

# CONCLUSION

The vinegar seeds analysed have a diverse microflora, including fungi that are potentially pathogenic when related to seed pathology, causing serious damage to agriculture. Therefore, there is a warning about the certification of seeds and their sanitary quality. In order to have good productivity when planting, it is necessary to invest in the quality of the initial product.

## REFERENCES


AL-WANDAWI, H.; AL-SHAIKLY; ABDUL-RAHMAN, M. Roselle seeds anew protein source, **Journal of Agricultural and Food Chemistry** 32(3): 510, 1984.

ASHRAF, H.; JAVAID. A. Evaluation of antifungal activity of *Meliaceae* family against *Macrophomina phaseolina.* **Journal of Mycology and Hytopathology,** Lahore, v. 5, n. 2, p. 81-84, 2007. Available at:<http://www.pu.edu.pk/mppl/journal/previousissue/Mycopath-3.pdf> Accessed on: 15/08/2014.

AKINDAHUNSI, A. A., OLALEYE, M. T. Toxicological investigation of aqueousmethanolic extract of the calyces of Hibiscus sabdariffa L. Journal of Ethnopharmacology 89, 161-164p. 2003.

BAKER, K.; SMITH, S. H. Dynamics of seed transmission of plant pathogens. **Annual Review of Phytophathology,** Paio Alto, 3:311-44, 1966

BAKER, K. Seed pathology - concepts and methods of control. **Journal of Seed Technology,** 4(2): 57-67, 1979.

BAKER, K. Seed pathology. In: KOZLOWSKI, T" ed. Seed biology. New York,

**Academic Press,** 1972. v. 2, p. 317-416.

BARROS, S. T. ; OLIVEIRA, N. T. ; MAIA, L. C. Efeito de extrato de bulbo de alho, (A//ium *sativum* L.) sobre o crescimento micelial e germinação de conídios de *A/ternaria* spp. E *Curvu/aria* spp. **Summa Phytopathologica**, Botucatu, v.2, n.2, p.60-62, 1995.

BETTIOL, W. **Biological control of plant diseases.** Jaguariúna:

Embrapa-CNPDA, 1991. 388p. Documents, 15.

BOTELHO, L. da S. **Fungi associated with semen of ipê-amarelo (Tabebuia serratifolia), ipê-roxo (Tabebuia impetiginosa), aroeira- pimenteira (Schinus terebinthifolius) and aroeira-salsa (Schinus molle): incidence, effects on germination, transmission to seedlings and control.** Dissertation (Master's) ESALQ, Piracicaba. 2006. p.114.

BRAZIL, Ministry of Agriculture, Livestock and Supply. **Manual of non-conventional vegetables.** Secretariat for Agricultural Development and Co-operativism. - Brasília: Mapa/ACS, 2010a. 92 p.

BRAZIL, Ministry of Agriculture, Livestock and Supply. **Manual of non-conventional vegetables.** Secretariat for Agricultural Development and Co-operativism. - Brasília: Mapa/ACS, 2010b. 92 p.

BRAZIL. Ministry of Agriculture, Livestock and Supply. **Manual for the Sanitary Analysis of Seeds.** Secretariat for Agricultural Defence. - Brasília: Mapa /ACS, 2009. 200 p.

CÁCERES, Armando et al. Antigonorrhoeal activity of plants used in Guatemala for the treatment of sexually transmitted diseases. J.

**Enthopharm.** Amsterdam, v.48, n.2, p. 85-88, 1995.

CAMPOS, F. M. et al. Pro-vitamins A in vegetables sold in the formal and informal market of Viçosa (MG), in three seasons of the year. **Ciên. Tecnol. Alimen.** v. 26, n. 1, Campinas, Jan/Mar. 2006.

CARDOSO, M. O. (Org.) **Non-conventional vegetables from the Amazon.** 1. ed. Brasília: Embrapa-SPI, 1997. 150 p.

CARNEIRO, S. M. T. P. G. Action of neem on phytopotogenic fungi. IN: **NIM - _Azadrachta indica:_ nature, multiple uses, production.** IAPAR: Londrina, p. 59 - 64, 2002.

COVENTRY, E.; ALLAN, E. J. Microbiological and Chemical analysis of nem (Azadirachta indica) extracts: New data on antimicrobial activity. Phytoparasitica, Rehovot, v. 9, n. 5, p. 441-450, 2001. at:<http://www.springerlink.com/content/t112666183100504/fulltext.pdf? page=1>Accessed: 14/08/2014.

COWAN, M. M. Plant Products as Antimicrobial Agents. **Clinicai microbiology Reviews,** Washington, v. 12, n. 4, p. 564-582, 1999. Available at:<http://cmr.asm.org/cgi/reprint/12/4/564> Accessed on: 14/08/2014. EMBRAPA Clima Temperado: Hibiscus: from ornamental to medicinal. 2011. Available at:<http://www.embrapa.com.br>. Accessed on: 05/11/2018.

EMBRAPA Florestas: The Cultivation of Neem for Fruit Production in Brazil.

Available at:https://www.infoteca.cnptia.embrapa.br/bitstream/doc/315642/1 circtec162.pdf >. Accessed on: 05/11/2018. Technical Circular. Colombo, PR, 2008.

FOOD INGREDIENTES BRASIL. Plant extracts. **Magazine-fi.** N° 11, 2010 p. 16-

20. Available at http://www.revista-fi.com/materiais/120.pdf Accessed on: 02/09/2014.

GOSMANN, G.; MELLO, J. C. P.; MENTZ, L. A.; PETROVICK, P. R. (Eds.). **Pharmacognosy: from plant to medicine.** Porto Alegre: UFRGS, 2001. p.633-650.

GOVINDACHARI, T. R. et al. Identification of antifungal compounds from the seed oil of *Azadirachta indica.* **Phytoparasitica,** Rehovot, v. 26, n. 2, p. 109-116, 1998.

GULERIA, S.; KUMAR, A. *Azadirachta indica* leaf extract induces resistance in sesame against Alternaria leaf spot disease. **Jounal of cell and MolecularBiology,** New York, v. 5, p. 81-86, 2006.

HAIKAL, N. Z, Improving biological control of *Fusarium* root-rot in cucumber *(Cucumis sativus* L.) by allelopathic plant extracts.
**International Journal ofAgriculture and Biology,** Faisalabad, v. 9, n. 3, 2007. Available at: http://www.fspublishers.org/ijab/past-issues/IJABVOL_9_NO_3/17.pdf Accessed on: 12/08/2014.

HASSANEIN, N. M. et al. Efficacy of leaf extracts of neem *(Azadirachta indica)* and chlnaberry *(Meiia azedrach)* against early blight and wilt

diseases oftomato. **Australian** Jounal **of Basic and Aplied Sciences,** Faisalabad, v. 2, n.3,p. 763-772, 2008.

HEINZMANN, B. M. Compounds containing sulphur. In: SIMÕES, C. M. O.; SCHENKEL, E. P.;

KHATOUNIAN, C. A. **Food production for domestic consumption in Paraná: characterisation and alternative crops.**

Londrina: IAPAR, 1994. 193p. illustrated (IAPAR. Circular, 81).

KOUL, O.; ISMAN, M. B.; KETKAR, C. M. Properties and uses of neem, Azadirachta indica. Canadian Journal of Botany, Ottawa, v.68, n.1, p.1-11, 1990.

LAZAROTTO, M. et al. Detection, transmission, pathogenicity and chemical control of fungi in millet seeds (Ceiba *speciosa).*
**Summa Phytopathologica,** v.36, n.2, p. 134-139, 2010a.

LAZAROTTO, M. et al. Detection, transmission, pathogenicity and chemical control of fungi in millet seeds (Ceiba *speciosa).*
**Summa Phytopathologica,** v.36, n.2, p. 134-139, 2010b.

LAZAROTTO, M. et al. Detection, transmission, pathogenicity and chemical control of fungi in millet seeds (Ceiba *speciosa).*
**Summa Phytopathologica,** v.36, n.2, p. 134-139, 2010c.

LAZAROTTO, M. et al. Detection, transmission, pathogenicity and chemical control of fungi in millet seeds (Ceiba *speciosa).*
**Summa Phytopathologica,** v.36, n.2, p. 134-139, 201 Od.

LOPES, et al. Antifungal evaluation of *Allium Sativum* L. extract in the control of fungi in *Anadenanthera Colubrina* seeds. **Journal of Biology and Pharmacy.** ISSN 1983-4209. v 06. n[0] 01.2011.

LOPEZ, L.C.; CHRISTENSEN, C.M. Effect of moisture content and temperature on invasion of stored com by *Aspergillus flavus.*
**Phytopathology,** 57:588-590, 1967.

MACHADO, J. C. **Seed pathology: fundamentals and applications.** Brasília: Ministry of Education; Lavras: ESAL/FAEPE, 1988. P 107.: il.

MACHADO, J. C. **Seed treatment for disease control.**
Minas Gerais: Lavras - LAPS/UFLA/FAEPE, 2000a. 138p. : il.

MACHADO, J. C. **Seed treatment for disease control.**
Minas Gerais: Lavras - LAPS/UFLA/FAEPE, 2000b. 138p. : il.

MACHADO, J. C. **Seed treatment for disease control.**
Minas Gerais: Lavras - LAPS/UFLA/FAEPE, 2000c. 138p. : il.

MACHADO, J. C. **Seed treatment for disease control.**
Minas Gerais: Lavras - LAPS/UFLA/FAEPE, 2000d. 138p. : il.

MARTINEZ, S. S. **O Nim -** *Azadirachta indica:* **nature, multiple uses, production.** IAPAR: Londrina, p. 59 - 64. 2002.

MARTINS, M.A. de S. **Vinagreira** *(Hibiscus sabdariffa),* **a little-known wealth.** São Luis: EMAPA. 1985. 12 p (EMAPA, Serie Documentos 6).

MENTEN, J.O.M. Seed pathogens: detection, damage and chemical control. São Paulo SP. **Ciba Agro.** 1995.

MOSLEM, M. A.; EL-KHOLIE. Effect of neem *(Azadirachta indica* A. Juss) seeds and leaves extract on some plant pathogenic fungi.

**Pakistan Journal of Biological Sciences,** Faisalabad, v. 12, n. 14, p. 1045-1048, 2009. Available at: <http://scialert.net/pdfs/pjbs/2009/1045 1048.pdf> Accessed on: 14/08/2014.

MOTOYAMA, M. et al. Induction of phytoalexins in soya and sorghum and fungitoxic effect of citrus extracts on *Colletotrichum lagenarium* and *Fusarium semitectum.* **Acta Scientiarum Agronomy,** Maringá, v 25 n. 2 p.491-496. 2003.

Brazil, 25 April 2008. Available at:

<http://www. periód icos. uem. br/ojs/index. php/ActaSciAgron/article/view/2 062/1 617 > Accessed on: 19/08/2014.

MONROY-ORTIZ, C.; CASTILLO-ESPANA, P. Plantas medicinales utilizadas en el estado de morelos. México: Uaem, 2007. 405p.

MUKHTAR, M.A. The effect of feeding rosella (Hibiscus sabdariffa) seed on broiler chicks performance. Research Journal Animal and Veterinary Science, v.2, p.21-23, 2007. Available at: . Accessed on: 03/11/2018.

NEVES, B. P. das, **Technical Circular No. 28 Cultivation and utilisation of the Indian neem** *Azadirachta indica* **A. Juss.** Goiânia: Embrapa-CNPAF- APA.1996.

ONKAR O. D. Damage caused by microorganisms during seed storage. **Revista Brasileira de Sementes,** Minas Gerais: Viçosa vol. 7, n°1, p. 139-146, 1985.

PADOLE L.; THAKKAR, P. Neem use and potential in agriculture. **Neem Foundation,** Mumbai, 2009. Available at: <http://www.neemfoundation.org/neem-articles/neem-in- organicfarming/agricultural-usepotential.pdf> Accessed on: 18 September 2014.

PINTO, N. F. J de A. Sanitary analysis in the production of seeds of major crops. In: ZAMBOLIN, I. **Seeds phytosanitary quality.** Viçosa: Federal University of Viçosa, 2005. p.295-332.

POLTRONIERI, T. P. S. et al. Department of Plant Pathology and Entomology, Institute of Biology, Rio de Janeiro: Federal Rural University of Rio de Janeiro. **Summa Phytopathol.** Botucatu, v. 38, n. 2, p. 167, 2012.

RANASINGHE, L.; JAYAWARDENA, B.; ABEYWICKRAMA, K. Fungicidal activity of essential oils of *Cinnamomum zeilanicum* (L.) and *Syzygium aromaticum* (L.) Merr et LM. Perry against crown rot anthracnose pathogens isolated from banana. **Letters in Applied Microbiology,** Cardiff, v.35, p.208-211,2002.

REGO, A.M. **Diseases caused by fungi in Cucurbits.** Informe Agropecuário 17:48-54. 1995.

RODRIGUES, E. A.; SCHWAN-ESTRADA, K. R. F.; STANGARLIN, J. R.; SCAPIM, C. A.; FIORI-TUTIDA, A. C. G. Potential of the medicinal plant *Omuium gratissimum* in the control of *Bipolaris sorokiniana* in wheat seeds. **Acta Scientiarum,** Maringá, v.28, n.2, p.213-220, 2006.

SALES, N.L. **Effect of fungal population and chemical treatment on the performance of seeds of Ipê-amarelo, Ipê-roxo and Barbatimão.** Dissertation (Master's Degree in Plant Pathology) UFLA, Federal University of Lavras, Lavras. 1992. 89p.

SARTORATO, A.; RAVA, C. A.; YOKOYAMA, M. Principais doenças e pragas do feijoeiro comum no Brasil. 3 ed. Goiania. EMGOPA/EMBRAPA- CNPAF, 1987. 5p. (Documents, 52).

SAXENA, R.C. Naturally occurring pesticides and their potential. In: L. W. Shemilt (ed.). Chemistry and World Food Supplies: The New Frontiers, Pergamon Press, Oxford, 664p, 1983.

SCHWAN-ESTRADA, K.R.F, STANGARLIN, J.R.; CRUZ, M.E.S. Use of plant extracts in the control of phytopathogenic fungi. **Revista Floresta,** v. 30, n° 1/2, p. 129-137, 2000.

SILVA, G. S. da; PEREIRA, A. L. Effect of incorporating neem leaves into the soil on the *Fusarium* x *Meloigogyne* complex. **Summa Phytopathologica,** Botucatu, v. 34, n. 4, p. 368-370, 2008.

SILVA, M.B. et al. Antimicrobial action of medicinal plant extracts on phytopathogenic fungal species of the *Colletotrichum* genus.
**Revista Brasileira de Plantas Medicinais,** Botucatu, v. 10, n. 3, p. 57-60, 2008.

VIANA, P. A.; PRATES, H. T.; RIBEIRO, E. de A. Use of aqueous neem extract to control *Spodoptera frugiperda* in maize. **Technical Circular 88.** Sete Lagoas, 2006.

WETZEL, M.M.V.S. **Storage fungi.** In: Soave, J. &Wetzel, M.M.V.S. (Eds.) Seed pathology. Campinas SP. Cargill Foundation. 1987. p. 260-275.

# I want morebooks!

Buy your books fast and straightforward online - at one of world's fastest growing online book stores! Environmentally sound due to Print-on-Demand technologies.

Buy your books online at
**www.morebooks.shop**

Kaufen Sie Ihre Bücher schnell und unkompliziert online – auf einer der am schnellsten wachsenden Buchhandelsplattformen weltweit! Dank Print-On-Demand umwelt- und ressourcenschonend produziert.

Bücher schneller online kaufen
**www.morebooks.shop**

Printed by Books on Demand GmbH, Norderstedt / Germany